Glaciers of Europe:

5. Northern Norway (Svartisen).
6. Southern Norway (Jotunheimen).
7. Alps (especially Aletsch glacier).

Glaciers of Asia:

8. Himalayas

Glaciers of Australasia:

9. New Zealand (Southern Alps).

Glaciers of Africa:

10. Mt Kilimanjaro and Mt Kenya have small glaciers.

FACTS ABOUT GLACIERS AND ICE SHEETS

About 6 billion square miles of the Earth's surface are under ice. However the vast majority of the ice is found in just two places: Antarctica and Greenland, where it merges into huge spreads of ice called ice sheets. Elsewhere glaciers are found mainly in high mountain regions, especially north of the Arctic Circle.

The world's longest glacier is the Lambert Glacier in Antarctica, over 35 mi wide and over 300 mi long. The fastest moving glacier measured was in Greenland. It moved at up to 75 ft a day. The thickest ice in the world is in Antarctica, where it has been measured as nearly 3 mi.

The biggest iceberg floating in the sea was found near Antarctica. It had broken away from an ice sheet and it measured 200 mi by 160 mi. This is bigger than Wales, UK the state of Maryland in the US, or nearly as big as Taiwan.

The tallest iceberg recorded was 500 ft high and seen floating off the coast of Greenland. Because only one ninth of the real thickness shows above the sea, this iceberg must have been about one mile thick over all!

During the Ice Age, ice sheets formed on many low-lying areas of continents and spread towards the equator until they eventually covered nearly a third of the world's land surface. They have only shrunk back in the last few thousand years and could return at any time!

Grolier Educational Corporation
SHERMAN TURNPIKE, DANBURY, CONNECTICUT 06816

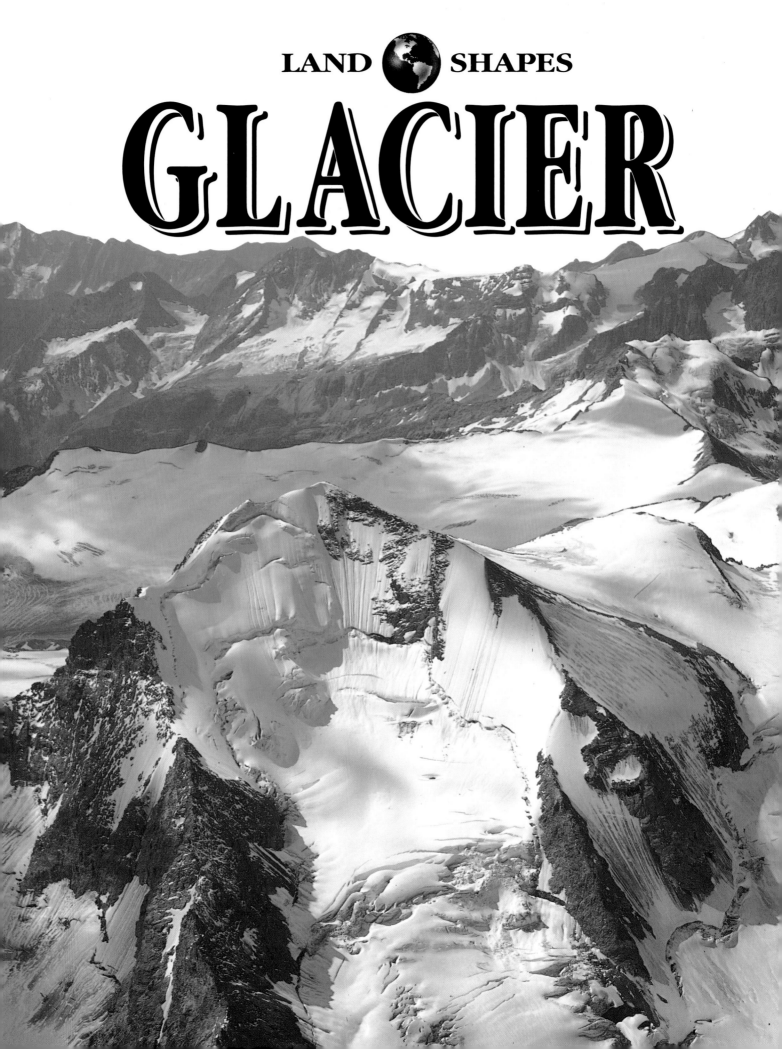

LAND 🌐 SHAPES
GLACIER

Author
Brian Knapp, BSc, PhD
Art Director
Duncan McCrae, BSc
Editor
Rita Owen
Illustrators
David Hardy and David Woodroffe
Print consultants
Landmark Production Consultants Ltd
Printed and bound in Hong Kong
Designed and produced by
EARTHSCAPE EDITIONS

First published in the USA in 1993 by
GROLIER EDUCATIONAL CORPORATION,
Sherman Turnpike, Danbury, CT 06816

Copyright © 1992
Atlantic Europe Publishing Company Limited

Library of Congress #92–072045

Cataloging information may be obtained
directly from Grolier Educational Corporation

Title ISBN 0–7172–7188–9

Set ISBN 0–7172–7176–5

Acknowledgements. The publishers would
like to thank the following: Horizon Air,
Peter Limieux of Athabasca Glacier Icewalks
and Redlands County Primary School.

Picture credits. All photographs from the
Earthscape Editions photographic library except
the following (t=top, b=bottom, l=left, r=right):
Nigel Bonner 32l; ZEFA 13t, 32/33.

Cover picture: Grindel Glacier, Grindelwald,
Switzerland.
Inside back cover picture: Bugaboo Glacier Alpine
Recreational Area, British Columbia, Canada.

In this book you will find some
words that have been shown in **bold**
type. There is a full explanation of
each of these words on page 36.

On some pages you will
find experiments that you
might like to try for
yourself. They have been
put in a blue box like this.

In this book mi means miles and
ft means feet.

People appear on a number of
pages to help you to know the
size of some landshapes.

CONTENTS

Introduction

Ice is one of nature's most powerful landshaping tools. At first this may seem strange, because the snowflakes of a snowstorm look as if they are some of nature's most fragile creations.

Ice covers one tenth of all the world's land surface – in fact it covers more of the world than the hot deserts. However, over the centuries the weather gradually changes. Even our distant ancestors knew what it was like for the land to be much colder, for some of them were living at a time called the **Ice Age**. Beginning 1 million years ago, the ice grew until it covered nearly a third of the world's land in vast blankets called **ice sheets**.

The ice sheets only began melting back a few tens of thousands of years ago and they have occupied their present positions for less than 10,000 years. This is why much of the world still bears the scars of the Ice Age.

Glaciers, which are like rivers of ice, are found in mountainous areas where they are trapped between peaks and ridges. When a new phase of the Ice Age begins, the glaciers will swell and cover the mountains with ice sheets once more.

Find out how powerful ice can be and what spectacular landshapes it can cut. And see how much material has been left behind to make new landshapes. Simply turn to a page and enjoy the landshapes ice has left for you to see.

This picture was taken inside a large crack, or crevasse, on the surface of a glacier.

Chapter 1:
Features of glaciers

What is a glacier?

A glacier is a tongue of ice that fills the bottom of a mountain valley or hollow. Its powerful effects shape the land dramatically as you can see in this picture.

Every winter the upper part of the glacier –often an **icefield** – is fed with fresh snow and compacted into ice. As the icefield gets thicker, so the ice is squeezed down the valley to make a glacier much like toothpaste being squeezed from a tube.

To find out more about the action of frost see the books Mountain *and* Valley *in the* Landshapes *set.*

As ice flows over rock surfaces it sometimes forms ice caves (see page 18).

How a glacier makes landshapes
As a glacier moves down a valley it carries pieces of rock that it has **eroded** from the valley floor and walls. When the glacier melts away the fragments are left behind as spreads of material called **moraine**. In this way glaciers make landshapes when they advance *and* when they retreat.

Icecaps and ice sheets are the sources of many glaciers (see page 14).

Mountain hollows are often home to small **cirque** glaciers (see page 24).

Most of the work of the ice goes unseen where the ice scrapes against the rock (see page 22).

Surface ice is brittle and cracks up to make **crevasses** (see page 16).

Rock fragments eroded by the glacier or shattered by frost collect on top of and underneath the ice and are called moraine (see page 28).

The making of glacier ice

A small block of glacier ice looks no different from the ice cubes you use to make cold drinks. But if you look closely you can see that glacier ice has lots of hairline cracks running through it. These tell you that it has been made under pressure.

Incredible though it may seem, hard glacier ice is made of soft snowflakes.

The weight of snow

How can tiny, almost weightless, snowflakes make a glacier that can tear and grind away at rock?

Freshly fallen snow is very light because the star-shaped crystals do not fit together very well. But the points of the crystals are extremely delicate and easily broken and crushed. This is what happens when snow falls upon snow during a snowstorm.

If you looked at deeply buried snow you would find that it was quite compact, with its crystals all crushed together. Snow crystals are also crushed when you walk on it or when you make a snowball.

Freshly gathered snow is already different from snowflakes because the flakes crush immediately when they reach the ground.

The snow crushed into a snowball. The more the snow is crushed, the harder the snowball becomes.

Although snowflakes all have the same basic set of six 'points', they form many patterns.

It is these points, or projections, that stop the flakes from fitting together.

Make ice

You can see how easily glacier ice forms if you make a snowball. While on a ski-holiday or in winter you may be able to try this using 'real' snow. At other times you can try this by scraping some ice crystals off the sides of a freezer.

Hold the ice crystals in a cupped hand. Cup the other hand and press your hands together hard to make a snowball.

Look closely and see how all the delicate edges of the snowflakes have been crushed and melted under the pressure.

Break the snowball open and you will find you really have an iceball. What are the differences in the properties of a handful of snow and a snowball?

Glacier ice forms by a similar process of weight and melting under pressure.

Crystals of ice taken from near the surface of a glacier.

Icefields, birthplace of glaciers

Icefields are large areas of thick ice where much more snow falls in winter than can melt away in summer. These great masses of ice create enough pressure to force ice at their edges to flow outwards to make glaciers.

Icefields that occur on mountain ranges are often called icecaps. The largest icefields almost engulf both the peaks and valleys of mountains over thousands of square miles. These are called ice sheets.

The foreground of the picture below shows an icefield in Greenland. Notice how there are few mountain peaks showing through. In the background you can see mountains and deep valleys. This area which is now free of ice gives you some idea of how much ice is stored in an icefield. The ice is eventually squeezed out of the bottom of the icefield.

Ice-filled valleys.

Understand the power of ice layers

Simple measurements of a snowball can help you to understand a lot about how powerful at landshaping ice can be. A glacier might be 1500 ft or more thick. So how much would it weigh?

For example, a snow 'cube', that has been squashed as when you make snowballs and is about 6 in thick will weigh about 2 oz. With the help of a grown-up find out how much a column of this kind of squashed snow would weigh if it were 3000 ft thick.

Your answer should help to show why ice can scour the rocks so easily as it moves across the landscape.

This is a hole that has been melted in a glacier by running water. Notice how the layers of ice can still be seen as rings of light and dark (dirty) blue. The hole is 30 ft deep and at the top about 3 ft wide.

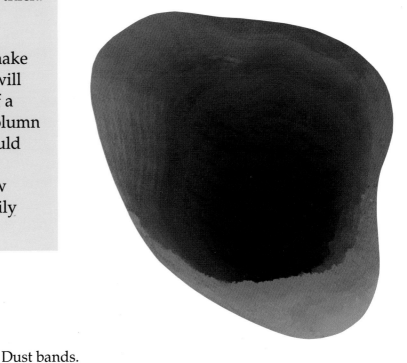

Dust bands.

Layer of snow.

This picture shows the layers of snow on a glacier. Each winter, snow falls on the glacier crushing the lower layers and changing the snow to dense ice.

Each layer here is separated by a dark band of dust blown on to the ice during each summer. How many years build-up can you see?

Crevasses

Glaciers move down hill at rates that vary between a few inches and tens of feet a year. Sometimes you can hear a glacier moving because it creaks and groans much like an old sailing ship in a storm.

Each creak or groan tells you that the surface has cracked open a little wider or that a part of the ice has slid a little further over the valley floor.

Why crevasses occur

A glacier slips forward more quickly in the middle of a valley than at the sides where it is scraping against the valley walls, or when the valley floor suddenly steepens.

This difference in movement causes the brittle ice to crack on its surface. When a glacier cracks open it makes a wedge-shaped chasm called a crevasse.

Crevasses are commonly found where the ice has to bend. Glaciers that flow swiftly from high mountain hollows are often covered with crevasses as the picture above shows.

This picture on the left shows a crevasse. The blue color is produced by the way sunlight is trapped by the ice. Only blue light passes through. The pressure gets much greater deeper in the ice and therefore the ice is forced together.

The glacier shown below bends around a curve and falls sharply over a rock step producing a mass of crevasses called an ice fall. There are so many crevasses that the surface is reduced to ridges. These ridges are called **seracs**.

Curve

Fall

17

The importance of ice caves

Glacier ice may be many thousands of feet thick, and at its base there is enough pressure to make the ice mould itself to the general shape of the valley floor.

But ice is not able to mould itself to every small change in the rock bed, and when there are rock steps, for example, the ice will have to arch over them. This produces caves.

When ice caves form at the side of a glacier it is possible for experienced people to go under the ice and see how the glacier is moving for themselves. This is what they find.

In just the same way as a river goes over a waterfall so a glacier sometimes goes over a rock step. The ice is brittle, so as it bends the surface pulls apart and cracks to give crevasses while the underside produces a series of caves.

Ice cave forms between the ice and the rock step.

Rocks are frozen to the underside of the ice.

Rocks are plucked from here.

Glacier ice moves this way.

Rock covered with ice.

The rocks are crushed between the ice and the valley floor.

Ice cave.

You can get an idea of size from the people standing inside this ice cave. In the picture you can see a frozen waterfall and also there is a sheet of ice across the cave floor.

The dark bands in the roof of the cave are pieces of rock that have become trapped in the ice. They will be used to scrape at the rock floor farther down the valley.

What an ice cave tells us

Inside an ice cave, such as the one in the pictures on this page, you can see a glacier at work. As the ice slips downslope it plucks at the rocks on the upper edge of the cave, tearing some of them away.

The plucked rocks are frozen tight to the ice and carried over the roof of the ice cave and down to where the ice meets the valley floor. Here the plucked rocks are scraped against the valley floor and finally crushed. The fragments of rock made this way are called moraine.

Scraping and crushing take place at thousands of places below the ice. These processes are the most important ways in which a glacier erodes its valley. They are so powerful that a glacier can erode its valley hundreds of times faster than a river.

Investigating how ice behaves

Understanding the way ice behaves is the key to understanding how a glacier shapes the land. Here are some investigations you can try to make the role of ice easier to appreciate.

Scratches make rock flour

The bottom of a glacier is full of debris. To see how debris scours the valley floor, put some sand or gravel in the bottom of an ice cube tray and then fill it with water and freeze it.

Take out an ice cube and push the cube over the surface of an old piece of plastic or an old roof tile. You will soon see the scratches it makes. But it also produces fine debris. In a glacier this fine material is called rock flour and it is soon swept up by the glacier as it passes.

Feeling the weight

To get some idea of the effect a thick glacier has on the rock below it, try this experiment.

Put one hand with the palm flat on a tray of sand and push it forward without pressing down. It is easy to do because there is little weight, but you don't move much sand either.

Now put your other hand on the back of the first one and press down a little while you push forward. Now your hand is more like a bulldozer. This is the effect ice has on any debris in its path.

Look for evidence of scouring

If you visit a mountainous area you will see many smooth, bare rocks. Many of these will have been shaped by ice scouring. Look at these to see if they have any signs of tell-tale scratch marks called **striations**.

A candy bar crevasse pattern!

This is the shape of a stone that has been scraped along the bed of the glacier.

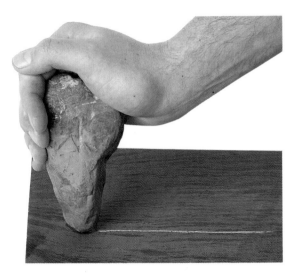

Natural banding of the rock.

Investigate crevasses

You can make a model glacier which gives realistic crevasses using some kinds of candy bar that have a chocolate coating, a crisp layer on one side and a caramel layer on the other. Here's how to enjoy the scientific experience!

a) Arching a glacier

Hold the candy bar flat with the crispy side uppermost. Look at it end on and imagine it as a glacier filling a valley. Now bend the 'glacier' to see what would happen when ice arches over a rock step. The surface should become crazed with crevasses.

b) Spreading ice

The lower end of a glacier is called the **snout**. Here glaciers often spread out, or **splay** out so crevasses appear. See how this works by squeezing the candy bar with your thumbs while keeping it level.

c) Melting away

Because the world's climate is slowly getting warmer, glaciers are melting away a little more each year. A glacier melts most noticeably at its snout and so it retreats a little more up its valley each year. You can imitate the warming of the world by eating the candy bar from the end up. Enjoy!

Scratch marks (striations) made on the rock surface as ice moved from left to right.

These are scratch marks made by a glacier on rock. They must have been the last ones the glacier ever made before it melted away at the end of the Ice Age.
The boulder has been placed on the rock to show the way it might have moved when the ice was present.

Chapter 2:
Landshapes made by glaciers

Glacial scenery

As glaciers erode the land they produce a unique scenery. Some ice forms in hollows high on mountainsides. Long rivers of ice gouge deep into the bottoms of valleys, and in some places ice actually spills from one valley to another, cutting right across high ridges. These features are each described in detail on the following pages, but each one is also shown in the picture below.

A pyramidal peak produced by several cirque glaciers eroding the sides of a mountain summit (see page 25).

An ice fall over the lip of the cirque hollow (see page 24).

A bowl-shaped hollow scoured by ice. It is called a cirque (see page 24).

A band of frost-shattered debris carried along the side of the ice. It is called a lateral moraine (see page 28).

A band of frost-shattered moraine on the glacier surface. It is called a medial moraine (see page 29).

Debris plucked and crushed by the ice and being carried under the ice. It is called ground moraine (see page 29).

Landscape spotting
Test yourself to make sure you
can see all the features marked
on the diagram below in this
photograph. You will find a large
version of this picture on the
inside rear covers of this book
to make spotting easier.
You should be able to find at least
one example of every surface
feature, but check with your
friends to see who can spot the
largest amount of each feature.

A small tributary
valley filled with
ice. It is called a
hanging valley
(see page 27).

Crevasses make a curved
pattern and show that the
middle of the glacier of moves
faster than the sides (see page 28).

A sharp-sided ridge
formed as two valleys
are widened by glaciers.
It is called an arete. (For
more information on aretes
see the book Mountain in
the Landshapes set.)

The U-shaped valley
profile (see page 26).

This ridge of
debris is called a
terminal moraine
(see page 30).

Glaciers in mountain hollows

One of the most sheltered places in a mountain landscape is in a hollow. During a blizzard a sheltered hollow quickly begins to fill with snow, and, with each new blizzard snow piles on snow to crush the lower layers to ice.

Size for size, ice is much heavier than snow, and when enough ice fills the hollow it begins to slip and slide, scouring out an easily recognised bowl-shaped hollow.

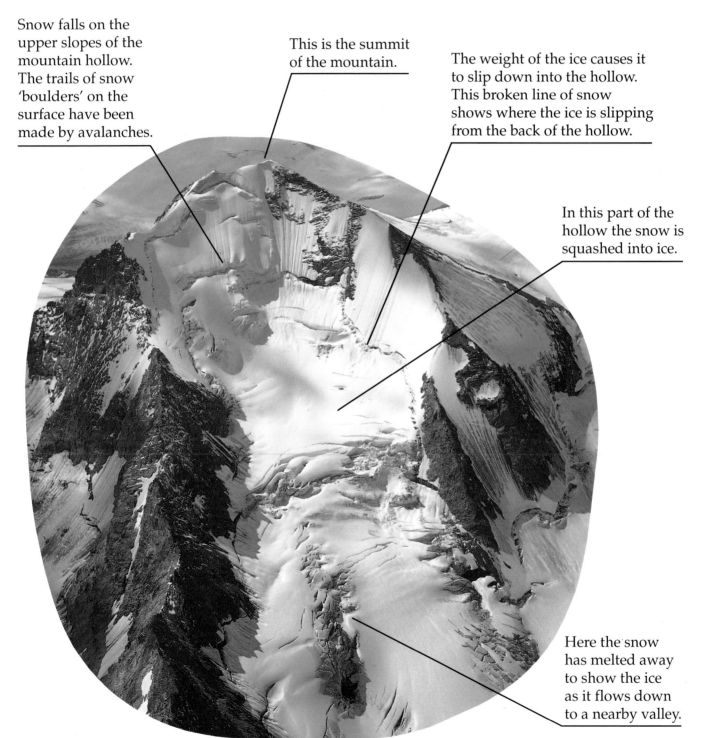

Snow falls on the upper slopes of the mountain hollow. The trails of snow 'boulders' on the surface have been made by avalanches.

This is the summit of the mountain.

The weight of the ice causes it to slip down into the hollow. This broken line of snow shows where the ice is slipping from the back of the hollow.

In this part of the hollow the snow is squashed into ice.

Here the snow has melted away to show the ice as it flows down to a nearby valley.

For more information on peaks and horns, see the book Mountain *in the Landshapes set.*

Three hollows have been formed on the flanks of this mountain, helping to sharpen the slopes leading to the peak. If many hollows form around a peak it will look like a steep-sided pyramid. This is called a horn.

This is the side of the main valley.

Here snow covers the ice.

In this lower region it is warmer and the snow has melted to show the ice below.

Ice-scoured hollows

As ice slips inside a mountain hollow, pieces of loose rock trapped underneath scour out the hollow into a shape like a wide open bowl. This bowl-shaped hollow has many local names such as **cirque** and corrie.

In the picture above you can see three cirques. Although no ice now reaches the main valley, during the Ice Age these cirques would have helped to 'feed' a valley glacier.

Make a cirque

Fill a saucer or shallow bowl with water and put in the freezer until it has become solid ice. Put a stick in it to act as a handle.

Make a slope in a tray of sand to represent the mountain face. Using the stick handle, push the ice into the slope to make a hollow. Keep turning the ice in the hollow, making strokes from the top to the bottom. This will soon make a mountain hollow and move debris from the top to the bottom of the 'glacier'.

U-shaped valleys

As a glacier flows in a valley, nearly all of its scouring power is concentrated on the bottom. As a result, glaciers deepen and widen valleys into trenches that sometimes look like a capital letter U. These U-shaped valleys are much straighter than river valleys because flowing ice cannot twist as easily as water.

These are small cirque glaciers just like the ones on page 24.

Valley glaciers are much larger and longer than cirque glaciers.

Notice how the U-shaped valley has very steep sides.

For more information about U-shaped valleys and hanging valleys see the book Valley *in the* Landshapes *set.*

Where did the glacier cut?

The picture above shows a valley after the ice has melted away. It has a clear U shape.

You can see how far up the valley sides the glacier used to be because the upper 'shoulders' are much more gently sloping. They were never gouged by glaciers.

Try to make a sketch of the valley and mark it where you think the ice came.

When you are on vacation or on trips, look to see if there are any straight, steep-sided valleys. This is a sure sign that glaciers once filled the valley.

What glaciers carry

Many glaciers have 'stripes' on their surfaces, just like the stripes on some toothpastes. It is these stripes that help you to see that the glacier is moving.

Glaciers carry immense amounts of shattered rock from the mountains towards the valleys. All rock fragments carried by glaciers are called moraine. They move much like they were being carried on an enormous conveyor belt.

Surface moraine

The edge of a glacier is often littered with a large number of rocks and stones. All of this material has fallen from the surrounding mountains and come to rest on the ice. As the ice moves along its valley this **lateral moraine** is also carried along. This is what the 'stripes' look like close up. Notice how this material has sharp edges.

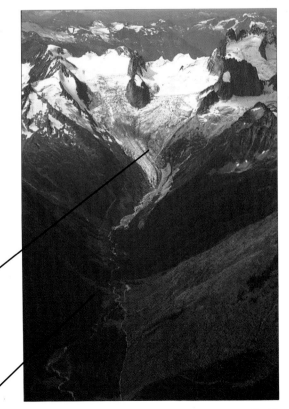

This is the
end, or
snout, of
the glacier.

This is the ground
moraine that has
been scoured from
the valley and carried
along under the ice.

Hidden moraine

The material on the surface of the glacier looks impressive, but it is only a tiny amount compared with the material carried and scraped along underneath a glacier.

This material – known as ground moraine – can be seen in ice caves (see page 20) and it finally emerges at the end of the glacier as sticky clay and boulders. It has rounded edges from being scraped against the valley floor, quite unlike the moraine that is carried on the surface of a glacier.

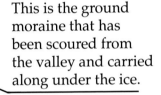

This dark line down the middle of the glacier shows where two stripes of lateral moraine have merged as two glaciers meet. It is called a **medial moraine**.

The ground and surface moraines are left behind as the glacier melts away. They weather to give soil on which trees can grow.

29

Discovering where glaciers have been

There is evidence of glaciers having covered much of the Earth, although it is not always easy to spot. Glaciers cut deep U-shaped valleys, and in doing so produce enormous amounts of broken rock. So a change in the landscape in the mountains is matched by a change in the lowlands – in places where the rock was finally dropped.

This is why you are as likely to see signs of past glaciers under a fertile plowed field as in a region of mountains.

This thick sheet of clay and rock is made of materials scoured from distant mountains. Use the trees as a scale.

Sometimes the material is left as a high ridge and is called a terminal moraine. More commonly it is spread over the land as an irregular sheet.

In a field look for slightly rounded boulders and scratch marks (see page 20).

It is often easy to spot a valley that has been cut deeper by a glacier because it will have a wide floor and steep sides. Glaciers often scour an uneven floor. Lakes commonly fill in the deeply scoured parts of such valleys.

Because small glaciers are not as fast at deepening their valleys as large ones, their valleys are left 'hanging' high on large valley sides. There are several hanging valleys on the left of this picture. This is also a common place to find waterfalls, so try spotting hanging valleys from their waterfalls.

For more information on how terminal moraines trap lakes, see the book Lake *in the Landshapes set.*

Chapter 3:
Some glaciers of the world

Antarctica

By far the largest ice sheet in the world covers almost the entire continent of Antarctica. Less than one hundredth of the continent shows as rock, and the ice sheets also stretch out from the land to cover millions of square miles of ocean.

Antarctica is now the coldest place on Earth, and although little snow falls (it is, in fact, a desert region), so little ice melts in the cold weather that the ice sheets continue to grow.

Although the surface of Antarctica looks nearly level, the ice sheets hide a land of high mountains and deep valleys.

Eleven penguins show the size of this iceberg.

Icebergs breaking away from an ice sheet.

Ice sheets and icebergs
The large ice sheets in Antarctica stretch right down to the oceans surrounding it and the edges of the ice sheets float on the ocean waters.

As the waters move beneath the ice, the edges break away, producing the huge islands of ice called icebergs.

The ice sheets of Antarctica are so thick that the ice is firmly wedged in the valleys and its lower layers do not move at all.

The landscape is in 'deep freeze' with almost no landshaping taking place.

A supply vessel for one of Antarctica's research stations comes near to one of the few areas of exposed rock, and gives the scale of these immense ice sheets.

Columbia Icefield

The Columbia Icefield lies in the southern part of Canada's Jasper National Park. It is typical of the small mountain icecaps that are found in many high mountains.

The icefield is ringed by mountains, and the trapped ice sends glacial fingers out in all directions to carve deep and dramatic valleys. Two of them are occupied by the Athabasca and Saskatchewan Glaciers, two of the most visited glaciers in the world.

This is the Saskatchewan Glacier with the icefield in the distance. The glacier sits low down in the valley because it is very small compared with a few thousand years ago.
At present the glacier is not scouring its valley very quickly.

The Columbia Icefield feeds the glacier.

This is a medial moraine, made from rock splinters released by frost from the valley slopes.

This part of the valley is covered with moraine which has not yet turned into soil.

The lake is trapped between ridges of terminal moraine recently dropped by the shrinking glacier.

The Athabasca Glacier is very impressive to visit. It is over 6 miles long and spills from the Columbia Icefield in a steep course with two major steps. In between the steps there is a flatter area, shown in the picture below.
 For those who want to be adventurous there are guided walks across the ice, while for the less active there are specially-constructed buses that drive over the ice on the flat part shown below. This is the only section that is crevasse-free.

New words

cirque
a bowl-shaped hollow scraped out of a mountainside by a small glacier. Other common names are corrie and cwm

crevasse
a wedge-shaped chasm in the surface of a glacier. Crevasses show that the ice is brittle and that it will crack near the surface, but lower down the pressure is greater and the ice flows, so crevasses cannot form

erode
the process of removing material from the land and carrying it to another place. Glaciers erode the land by scraping and plucking as they slip over the land

glacier
a tongue of ice that fits into a valley or a mountainside hollow. A much larger body of ice would be called an icecap or an ice sheet

ground moraine
the debris carried underneath a glacier. It has been plucked from the valley floor

Ice Age
the time, from a million years ago, when the world's climate became colder and ice sheets formed on lowland beyond the mountains on all the northern continents. At the maximum of the Ice Age ice sheets covered nearly a third of all the land in the world. The ice sheets only melted away a few thousand years ago, which is the reason so many scoured landscapes still show the effects of glaciers so clearly

icefield
a large body of ice, usually high up in the mountains. Icefields supply the ice for many valley glaciers

ice sheet
a large body of ice, often covering lowland but also completely burying mountains, which occurs only in the world's coldest places such as Antarctica and Greenland

lateral moraine
the name for debris that falls from mountainsides and comes to rest on the edges of a glacier

medial moraine
a ridge of surface moraine that has been made by the merging of two lateral moraines

moraine
a deposit carried by a glacier frozen inside the ice, or on its surface and then left behind when the ice finally melted

serac
a knife-edged ridge that forms between two crevasses in places where the ice surface is very disturbed

snout
the end of a glacier or ice sheet and usually the place where the ice melts away

splay
the spreading out shape that occurs near the end of a valley glacier especially if the ice pushes out on to lowland where there is no valley to hold it in a narrow tongue

striation
a scrape-mark made on the surface of rock where glaciers have recently been

Index